BEHAVIOURS
RELATIONS
&
INTERHUMANITY

An Interdisciplinary Theory

On Fractally Healthy
and Conscious Behaviours

Stefan Alldén

Cover/Illustration — Lisa C Larsson, Tecknarlisa
Graphic form and typesetting — Stefan Alldén
Author — Stefan Alldén

Publisher: YoumanifY, Bengts Land 19, 443 73 Gråbo, Sweden
www.youmanify.se info@youmanify.se
Printing: Libri Plureos GmbH, Friedensallee 273, 22763 Hamburg, Germany. Purchases outside the EU may involve a different printer.

ISBN: 978-91-531-3403-9

Table of contents

~ Preface ~

Aristotle supposedly said, that the truth lies between two concepts. Despite that, when I studied philosophy at the University, they talked about the truth as if it was conceptual. For me, over time, truth has been increasingly displaced beyond the realm of concepts. It gives me the feeling of understanding rather than knowing. It is as if the enchantment fades when it is dressed in words. Yet, in our very paradoxical world, I get the sense that the truth seeks to be conveyed through the words.

And behind what this book seeks to articulate, giving it a place of honor in every person's soul and heart, this perpetually conceptless world dances, and seems to want to whisper; truth is what we so often have come to call love.

Thus, I have a peculiar feeling when writing this, while at the same time being convinced, that the low-key silence beyond concepts, is the language of consciousness and that everything else, are more or less subtle or grossly displaced translations of this silence. When I then take in the silence and believe it is so important that I want to put words onto it in order to convey to others, it becomes a constantly ongoing wrestling match between the words and the humility. And I can only hope that the words and the concepts I have placed upon the silence in this book, represents the truth as closely as possible. When all is said and done, I can only reconcile myself to my wish beyond concepts - that all people may live exactly as we are created to be.

Stefan Alldén

Lerum Sweden, December 2024

~ Introduction ~

The theory represented in this book has been tested with successful results in practical work with actual people in therapy on an individual level, as well as in psychosocial tutorials and lectures. The author welcomes further studies and collaborative research in this area.

It is intended to be used as course literature. The book is also applicable as a basis for psychosocial and group dynamic processes within companies, municipalities, institutions and other social structures focusing on community. In addition, it can serve as a basis for lectures, education and courses - together with or in collaboration with the author.

The book innovatively defines behaviours and health with fractal symmetry. It also includes a new definition of consciousness as a foundation. A new model of how behaviours arise, as well as a challenging mental model of how information is communicated is introduced. These are part of and also reinforce the foundation of the theory and help to concretize what we cannot see with the eye, but which exists and can be felt.

It provides new perspectives on consciousness and the prevailing attachment and behavioural theories. The author is of the opinion that these theories, with necessity, need to be renewed so that people can fully understand their behaviours and relationships and to be able to live healthier lives together with each other.

Readers are provided with short, concise and pithy explanations and the chapters which are intimately interwoven, give a gradually deepened understanding. Selected chapters then build on (self)-reflection and discussion.

Based on the concept of interhumanity, the book wants to take a closer look at what goes on between us humans. What it is, how to describe it and how it affects us. Thereby, it treats behaviours based on the structure of our bodily functions as a whole and how we receive the information that is the basis for our behaviours are based on. Relationships, male and female, are some of the common threads of the theory and are emphasized throughout along with what favours the most healthy.

In the same way that we can observe the symbiosis in the interactions in nature, the chapters of the book are meant to be connected. And this theory gives us a hopeful perspective on how we, with fairly simple means, can influence our world in a healthy way. While at the same time feel better ourselves, because on a deeper level, we will understand, that the key to health and well-being, lies in doing things for each other and focusing on the desirable.

* * * * * *

1

~ Embryological order of priority ~

If we are to go to the bottom of what behaviours are and how they arise, we need to **understand certain principles for how life begins at the fractal level** (see also chapter 2 Fractals). The reproduction is based on giving something when something is ready to receive it. And that what is given is accepted by the recipient (the latter is further developed, with what is relevant to human behaviour, in chapter 20 Male and Female).

Everything is propagated by external influence, ie something is given to be received, which generates a new life in some form. Human reproduction takes off in the fact that a specific sperm is accepted and received by the egg, where it is taken care of and fractally multiplied until a new human being is ready to be given back to the world (important to understand that the sperm **is accepted and received**). Inside the uterus, the embryo gets everything it needs to be further developed, from the outside. That is, from the mother who carries the embryo. The mother in turn feeds on what comes to her from outside to be able to feed her embryo. In this simple reasoning, we can understand how everything that arises, receives its information from the fact that external factors are accepted at each level (see also chapter 9 Sources of information and chapter 15 Behaviours are collective).

As far as fertilization is concerned, we can advantageously think in terms of cellular communication for the fractal understanding. Where **the outer casing of the cell protects the nucleus of the cell against unhealthy frequencies and permits the healthy ones to enter**. A fractally healthy cell then emits what is also fractally healthy for other cells and the embryo can continue to develop.

In the initial stage, one can say that the whole embryo is a heart, a cardiac stage, which gets its oxygen from the mother through her breathing and into the blood which is then linked to the cardiac stage. This cardiac stage, through which the embryo gets its oxygen, can thus be seen as a precursor to our senses. Note that in other words the heart's function is secondary breathing (cf. the principle of cardiovascular rescue). We can also verify this at birth, where breathing must get started with the newborn in order for life to continue (see also chapter 29 Heart, emotion, brain).

After the embryo's cardiac stage, the gastrointestinal tract is formed, which also is our emotional center. This part is thus prioritized as number two in the process, which also means that the external signals need to be verified by something inside to be permitted to enter (cf. the egg's reception of the sperm). And without going further into this, we can also understand that the embryo needs receptors (senses) to receive external signals (nutrition/information) and then emotionally determine whether it benefits the development or not (cf. outer casing/cell nucleus there the outer casing allows what benefits the nucleus of the cell).

After that, the rest of the body develops and last in the chain comes the brain. This does not mean that the brain is least important, but rather that it is very important for the interpretation of bodily signals through this order of priority (see also chapter 29 Heart, emotion, brain). If, on the other hand, we try to use the brain inverted (conversely) against the natural order of priority, we will encounter dilemmas. We deal with this in chapters 14 and 22 - The chain of behaviour and Learning. For the extension of this understanding, we should also note that the brain is an organ and not a sense. And that the brain is designed to receive bodily signals and then pass on the information further to determine what then will be the best action.

A simple example of this is when we burn our hand on a hot plate on the stove. The signal from the hand goes to the brain, which immediately sends signals to pull the hand away. And we have had a new experience based experience programmed into the fractal chain for future use.

An exciting interpretation of the initial part of conception and relevant to this context is when the sperm just entered the egg. In terms of space, in relation to the size of the sperm, it is like entering a giant hall where the sperm is taken care of and embraced by the egg and they both become completely still for the first 1-2 hours. A process of inclusion to examine if they can feel safe with each other. If they do, a new fractal symmetrical process begins, the one we call life, with a movement that can be likened to a first breath. That sets the electromagnetic template for the first fractal in that human fractal symmetry.

2

~ Fractals ~

Fractals can be explained as a pattern with **a detailed structure that is uniformly repeated at all levels and in all dimensions in a built-up life form** (cf. the cell division principle). We can call the structure of fractals for zoom symmetry. That is, what we can see with the naked eye, we can see in the same pattern at all levels if we zoom in the same with any aid. This pattern is the foundation for different complex life forms in our world. Man is part of it and is built up with the **fractal principle**.

The concepts of fractals and zoom symmetry we can then put together into the concept of **fractal symmetry**. Which will have a summary meaning that a fractal, in addition to being able to zoom in and see other similar forms at all levels, also has the character of being symmetrical in its chain of growing when creating different life forms. That is, it is copied symmetrically if it is not exposed to defects.

Example of a fractal symmetry

When we talk about fractals in this context, we must remember that we as human beings, have come to try to create images in front of us of what we are trying to understand. But, through the reading in the book, we will also delve into the comprehension that there is also something beyond these images. And in this chapter on fractals, we need to add the understanding that what is beyond the visible (images) is made up of electromagnetic fields. These electromagnetic fields are the innermost essence of the fractal symmetry. Let's remember this to bring along as a complementary understanding of the upcoming chapters.

What is created in the initial fractal symmetry will be copied and be a part of the entire life form when completed. For man, this means that the differences that we humans possess are dependent on what has been internalized in the fractal symmetry in each specific individual, from the fetal stage onwards. Both in terms of the physical (visible) as well as the psychic (invisible).

If a child develops unhealthy behaviours, it does not mean that there was anything wrong with the surroundings in any judgmental sense, but that some external influence, which does not comply with healthy fractal principles, has been internalized in the fractal symmetry and caused a defect from that level on in the **fractal chain**.

Physical defects in the **fractal structure** often manifest themselves so that we can see them on or in the body with or without aids. The psychic turns out in a different way, for example through what the book refers to as emotional blockages (see chapter 16 Emotions). These arise through psychological influences from the surroundings (electro-magnetic fields) (cf. chapter 1 Embryological order of priority) which creates trauma on electromagnetic levels (see chapter 3 Consciousness, chapter 11 Micro-trauma). These are encapsulated by the child and create emotional blockages that partly give rise to bodily tension and partly to fractal micro-trauma based behaviours later in life if no processing was performed. (See chapter 11 Micro-trauma).

Fractal symmetry with an unhealthy internalized structure, can be healed through inverted fractal processing (the back way, so to speak) and healing trauma on each level, deeper and deeper into the fractal structure. The entire blockage is gone when all the symmetry in the fractal chain has been repaired/ restored.

The fractal structure/symmetry is in itself a rewarding, collaborative, inclusive, curious and health-promoting function that makes up all of life. This in itself gives us indications that man in his essence has these parts built in for her purpose and to be manifested throughout the chain of behaviour (see chapter 14). It also means that it is healthy to follow the rewarding, collaborative, inclusive, curious and health-promoting fractal human structure. You can also metaphorically look at the fractal symmetry as a competition with only winners, where each healthy extension of the fractal chain leads to a strengthened structure for both the individual and its surroundings.

Figure 1. Romanesco broccoli

Figure 2. Fern leaf

Figures 1 and 2 shows examples of visible fractal structures in nature.

* Discuss/contemplate how to understand the fractal symmetry in the human body.

3

~ Consciousness - What exists, but is not visible ~

The definition of **consciousness** in this book reads - **what exists, but is not visible and is involved in creating what is visible and physically can be experienced through the senses**. This chapter provides a basic insight into this invisible world by concretising the abstract through our senses, emotions, thoughts and further on in the chain of behaviour according to that chapter. Please note that the concept, consciousness, differs from how the book defines awareness in the next chapter. But, since all chapter in this book are woven together, these two concepts are also fractally linked.

One way to take off in the realm of the invisible, is to describe it as energies, frequencies and vibrations via varying electromagnetic fields. These reach man through his senses. Visual impressions as well as experiences of sound and fragrance can be directly derived from this explanation, while taste and feeling often comes from food or touch. In these cases, we must remember that both food and touch are originated in the invisible field in their fractal processes (see also chapter 2 Fractals).

These electromagnetic fields are invisible to the naked eye but can via conscious presence be felt through our senses. As we learn to "read" these signals, the more the communication beyond concepts can take place. That is, being more aware of what is happening in our bodies (what is invisibly received and what is stored in the body) and what it means beyond words. Where consciousness (what exists, but is not visible) is available. In the chapter on Sources of information, this understanding will be expanded.

Being intellectually conscious (through learning - reading scriptures and learning from digital devices) is not the same as being emotionally conscious. Access to consciousness is only possible through our bodies, not through the brain which is a station for interpreting bodily signals (see also chapter 1 Embryological order of priority and chapter 22 Learning).

The goal of consciousness is for us to be able to feel each other and to get the emotional center and the brain in a congruent state. A state where emotions and the brain have access to information from the same source (senses). See also the next chapter, Awareness and unawareness.

4

~ Awareness and unawareness ~

Chapter 3 on consciousness ends with: *"The goal of consciousness is for us to be able to feel each other and to get the emotional center and the brain in a congruent state. A state where emotions and the brain have access to information from the same source"*. If the information to the brain and emotional center comes from different sources, we are in a non-congruent state. Herein lies the difference between being conscious and unconscious in this book's definitions of these concepts.

Not infrequently, the senses and emotions (the body) and the brain (the intellect) give us conflicting information, sometimes clearly and obvious and sometimes more subtle and on a hidden level. When we are in a state where what we experience intellectually does not agree with what we experience mentally and emotionally, then we are in a non-congruent state that should be examined in order to get clarity on what really is true.

In other words, we can express it as **intellectual information that is not (fully) anchored in the senses and emotional center (the body) means we are unconscious to some extent,** according to this definition. This also aligns with what the flow of information looks like according to the embryological order of priority and the fractal principle. **If the intellect (the brain) is provided with information from the senses and emotional center, then we are conscious to some extent** according to this definition. We become fully aware when all information follows the flow according to the embryological order of priority and the fractal symmetry remains intact, or when all emotional blockages (see chapter 16 Emotions) are processed and the fractal symmetry is restored.

The more accumulated experiences in our emotional life (experiences that have given rise to defects in or reinforcements of the fractal symmetry) as humans in different situations, the more information we possess. But, it only reaches our full consciousness if we are given the opportunity to process the emotional blockages that have arisen. Opposing feelings are important for the overall information. For example, it is not enough to have been bullied to understand what bullying entails. You must also have been the bully. If there is no processing, the information from emotional blockages remains consciously inaccessible to the brain. For each emotional blockage we process, information is released to the brain and the congruence between the emotional center and the brain gradually increases.

Below are two examples from the order of information; senses —> emotion —> thought; to help clarify the definition of the terms:

You have read everything about how post-traumatic stress works and can explain it in detail in writing and speaking. But, you haven't processed the emotional blockages from your own post-traumatic experiences. In this case, you are unconscious by the definition of the book, because you have only activated the thought and not accessed the feeling that contains the information your senses once gave you when the trauma was imprinted. The embryological order of priority and the fractal principle are not followed in this example.

In the second case, you have only processed your traumas and not read anything about them. But, you have accessed the emotional blockages and endured the pain/suffering that occurs during processing. Then the information is released and the brain receives experience based information from the sensory experiences when the trauma occurred, through emotion to thought. This is in harmony with the embryological order of priority and the fractal principle, which gives awareness according to this definition.

(cf. chapter 22 Learning)

Note of extra importance: If the information we receive does not agree with our senses, we need to question the information. The senses always tell the truth. But, we can believe that intellectual information is compatible with current sensory experience, because a certain type of intellectual information has become so entrenched in the world we live in, by many people believing in it (cf. cognitive dissonance).

5

~ Interhumanity ~

Interhumanity is the environment between us as humans (where we are). What the book terms energies, frequencies and vibrations via shifting electromagnetic fields, is at its best, when collaboration between individuals gives rise to emergence. In this, let us call it a gap, we emit and receive these fields (see chapter 14 The chain of behaviour). And since interhumanity is part of what we emit to others, interhumanity also becomes part of our common inner environment. Interhumanity is in principal, in this theory synonymous with consciousness.

6

~ Focus ~

A short example to start from: If I am determined to buy grass-fed meat from a farm store, I will resolutely go there, buy meat and bring it home.

This example implies that **what we focus on creates our reality (will be manifested)**. If the conscious part of the focus is greater than the unconscious part. However, I could have had an (unconscious) trauma that made it difficult for me to visit a farm shop, which was greater than the conscious part, that would make me cancel the trip. (see also chapter 11 Micro-trauma, chapter 4 Awareness and unawareness). Then that part had become my unconscious focus. Focus as a concept and definition in this book means that **the dominant focus manifests our (conscious) reality**. Based on that, consciously or unconsciously, we have a choice in every given opportunity (see also chapter 8 Healthy/unhealthy behaviours).

A quantum physics experiment called ”The double slit experiment” demonstrated that the observer's expectations of the outcome of the experiment affected the outcome of the experiment. That is, their focus on what they thought would happen, changed the behaviour pattern of the molecules to become what the observers expected. When the experiment was performed without their observation and expectation, the behaviour pattern was unaffected and the molecules could form their own pattern.

A more everyday way to describe that experiment, is by saying that the focus does not care about the word ”not”. So instead of e.g. saying; beware lest you fall, it is better to say; make sure to stay on your feet. The reason becomes obvious that in the former case we create an unwanted focus on falling. The latter instead creates the desirable focus to stay on your feet.

A funny simple thought experiment can exemplify how easy it is to create an unwanted focus - Don't think of a pink elephant!

A logical conclusion as a result of this, along with the other principles in this book, leads us into a deeper understanding of how the behaviours are influenced by how interpret and behave with each other. We can all relate to what it would feel like to be ignored, be left out of context or being bullied. Or for that matter, understand the difference between receiving or giving harsh or kind words and how we are affected by it. This makes it easy to understand what is most healthy to focus on.

It is also the case that we to the same degree are affected ourselves by what we focus on and what feelings we generate in others. A friendly, inclusive and positive focus on each other is always synonymous with health.

7

~ Presence ~

Presence is always where the body is. That is, where the senses have access to the flow of information from the environment according to how we are constituted. We can always imagine that we are somewhere else with help from some old memory (see chapter 23 Memory), but then we are not present to where the senses are at the moment. We need presence both when we have to access and process emotional blockages and when we are getting information directly from the environment around the body.

The brain can distance us from presence, for example, by a memory or trauma from the past coming up and getting us to a different place than where the body is. The brain also generates thoughts from the presence through where the body and senses are located. But, the brain cannot generate information from the presence by itself as it is an organ and not a sense. In this context, it is only intended to be used as a station for interpretation for bodily signals generated by presence-based impressions from the senses. This also applies to the impressions from the senses that are stored in the body from earlier in our life.

For healthy behaviours, presence is always beneficial. In the field of relationships, interhumanity, we need to be in the same flow of information in order for the communication to be fruitful. It can of course be at the same place based on a shared memory, but here we focus on how we can follow the embryological order of priority and the fractal principle for the right flow of information in presence. When we talk about a shared memory, this is followed in principle, albeit from another time (another level in the fractal symmetry), but the thought is not where the senses are right now.

Example: It is difficult to get the feeling of a sunny and warm day if we at the same time are outdoors in a torrential rain at +10 degree Celsius.

Communicative challenges occur when we are at different places as described above. Then we do not share the same flow of information with each other and we find ourselves in an unhealthy communication, distanced from presence as our basis for the flow in the chain of behaviour.

Breathing is a good practice for staying present if we follow it emotionally on its way into the body and then out again. It can also be used as a metaphor for how something from outside enters the body and is a prerequisite for presence.

8

~ Healthy/unhealthy behaviours ~

First, the definitions:

All behaviours that unite, create security, strengthen relationships and strive to see and do something for others - are healthy.

All behaviours that separates, creates insecurity, deteriorates relationships and strives to make others only see oneself and do what only benefit oneself - are unhealthy.

The embryological process, the fractal symmetry and the chain of behaviour verifies what is healthy and unhealthy through their respective structures, according to what we covered earlier. Which is confirmed and clarified in the definitions above. Also remember, that the behaviours described according to the definitions above, refer to all levels in the chain of behaviour. These are dealt with in chapter 14.

But, do we really choose between healthy and unhealthy behaviours? Here, we must bear in mind that based on given conditions in the fractal symmetry and the current situation, we do not in principle have a choice based on the teachings of this theory. But, it can be expressed as that depending on how strong the congruence between the senses/emotions and brain is, we can make more and more conscious choices. The more aware we are, the more healthy and cooperative behaviours can be expressed. Compare with chapters 4 and 6.

Suggested exploration: What is one's conscious healthy intention and what is one's unconscious unhealthy intention?

* Reflect and discuss.

9

~ Sources of information ~

When we can see our bodies as receivers and transmitters of information trough our senses and interacting (cellular) electromagnetic fields, we can also understand that we can generate different information depending on the actual environment (see also chapter 14 The chain of behaviour). The frequency or frequencies that the body is tuned to (fractally symmetrically programmed with), are the frequencies we will be able to receive (cf. chapter 6 Focus). The principle of how a radio works is applicable for a simple fundamental explanation. We set the radio to a certain frequency and then that is the frequency we can listen to.

Sources of information are basically everywhere. The question we should ask ourselves; which sources of information are healthy versus unhealthy respectively? Below is a list of how to think about what different sources of information are:

Media, advertising, the internet, school, academia, curriculum, authorities, institutions, laws, rules, regulations, libraries, publications, cultures, art, religions, wireless technology, artificial intelligence, music, sound in general, noise, chemicals, stimulants, medication, food, cosmology, nature, climate, sun, wind, water, vegetation, photosynthesis, animal life, social context, other people, breathing, etc.

We can then categorize these as healthy or unhealthy. That is, which environments/frequencies lead cells and people being either more cooperative or more competitive with each other. Where cooperation always should be seen as healthier than competition (see chapter 26 Cooperation and competition).

We will not go further into which sources generates health versus unhealthiness, but again just briefly mention the principle that what harmonizes with the inner resonance of the nucleus of the cell and the man respectively, is healthy (see also chapter1 Embryological order of priority 3rd paragraph).

We also pose the questions: Why are there unhealthy sources of information? And: Would defects in fractal symmetry heal themselves and return to their origin if only healthy sources of information existed?

* Contemplation, reflection (see chapter 17 Self-reflection) and discussion.

10

~ Attachment ~

In this book, the concept of attachment is misleading because it traditionally refers to the child trying to find strategies to bond with their parents. Another way to express it, in accordance to the concept of this book, would be to say that the child wants (demands) something from its parents (because they could not give on all levels due to a defect in the fractal symmetry, see chapter 11 Micro-trauma). That contradicts our inner fractal structure of giving and results in the development of unhealthy behaviours and strategies for relationships (see also chapter 27 Codependency).

To understand what this book intends with the concept of attachment in a fractal context we need to use some other words. So, instead of attachment we call it; **a quest for staying in the origin of the healthy fractal symmetry**. What the child then does, is try to stay in its original healthy state in an unhealthy environment, ie the defective ability of the environment to communicate (give unconditional love) on all levels, which will not be successful. Instead, the child closes off feeling by feeling (encapsulates and they become emotional blockages, see chapter 16 Emotions) because the communication on the level of the senses and emotions does not work as intended. This environmental insecurity, successively leads to unhealthy separation of emotional centers/brain and subsequently does not follow the fractal chain and the embryological order of priority.

Later on in this book, this comprehension can be linked with the content of the chapter on male and female. A boy who experiences changes in his fractal symmetry during the early years of life will need processing in order to restore to his original male structure. And the same applies to a girl as far as her female structure is concerned later in life.

As stated, attachment according to this theory is based on; *"a quest for staying in the origin of the healthy fractal symmetry"*. In intact fractal environments, the child will be successful because the child only gets healthy information through the chain of behaviour.

11

~ Micro-trauma ~

The term m**icro-trauma** is here used to describe what happens when we get **an emotional blockage that differentiate the brain and the senses. That is, there is a discrepancy between the intellectual information and the information from the senses & the emotions**. The little child who wants to communicate with emotions from impressions of the senses (and can only initially), on such occasions, micro-trauma is created by the child not reaching its surroundings within the basic fractal symmetrical communication. The environment gives an incorrect intellectual input of what they think the child means and the child does not reach out to convey its message and micro-trauma arises.

For instance, when a child is screaming and the parents think the child is hungry, but in fact the child wanted emotional closeness - micro-trauma arises. Or if a child points at a lamp in the ceiling because the sensory impression of the light creates a certain feeling and the child wants to share but the parents respond saying that it's a lamp. The emotional information from the child is converted into intellectual information about the name of the item that the child points at - micro-trauma arises. Over time, the child gives up their attempts to communicate emotionally in most situations. Instead it encapsulatea feeling by feeling and tries to reach its surroundings intellectually and gradually takes on the language to communicate that way instead.

From this we can understand that there are many micro-traumas during the first years of life and all of these interact in situations when one of these traumas is recalled later in life. If the synergistic effect of the micro-trauma is greater than the effect of emotional impressions from the actual moment/presence, this will dominate and create an unhealthy behaviour/

unhealthy communication. The bodily functions that are actuated at this point, are what we call fight or flight (see also chapter 18 Stress).

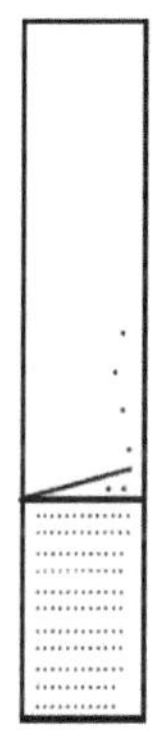

Figure 3. Column as a symbol of the body.

To concretize and symbolize this reasoning, we can use the column in figure 3 above. The column symbolizes a human body where the dots are micro-traumas. The body (column) is completely clean and free from micro-trauma at birth (no dots in the column). Here, we can ignore the philosophical reasoning of whether or not we are born as a blank slate and the possible defect in fractal symmetry that may have been caused during pregnancy or from the environment during pregnancy.

The dots in the lower part of the column thus symbolize the micro-traumas and the size of that field depends on the factors of how the child was raised (see also chapter 10 Attachment). This field interacts as mentioned above and lies latent as fractal symmetric defects and it affects our behaviours when active. The slanted line which releases some of the micro-traumas, symbolizes what happens when processing micro-traumas. This means that emotional blockages are dissolved, information is released, fractal symmetry is repaired, trauma is brought to awareness and the baggage that blocks the way of the healthy presence is relieved (see also chapter 31 Process and manage).

Note: The micro-trauma manifest themselves differently in men and women at the times they are activated and affect behaviours later in life (see also chapters 16, 20, 31).

12

~ What is a behaviour? ~

The answer derives from the fractal symmetry, which is basically a chain of motions (see chapter 2 Fractals). And **a behaviour is a movement** at all levels with reference to the chain of behaviour. Our senses receive electromagnetic movements. The senses pass on these movements to the dynamic emotional life which in turn generates a thought. Then the movement is passed on to speech, physical action, a habit and will finally become our character.

With that reasoning, a behaviour becomes something that cannot be isolated. No behaviour can therefore be individual. The healthy conclusion is that no one can do anything wrong, because everyone just did what was internalized in the fractal symmetry. The act of self-reflection is what fertilises the healthy repair of the fractal symmetry.

13

~ How does a behaviour arise? ~

When all the cooperating conditions for a certain behaviour take place - then the specific behaviour occurs. The fractal structure makes no mistakes. But if needed, it can be repaired with the right help/tools (see also chapter 4 Awareness and unawareness, chapter 16 Emotions, chapter 2 Fractals and chapter 31 Process and manage).

In other words; behaviours arise from external interacting stimuli, together with internal interacting, in the fractal symmetry internalized stimuli. Behaviours are thus always a response to stimuli. Where stimuli are synonymous with presence based sensory impressions (see chapter 7 Presence).

14

~ The chain of behaviour ~

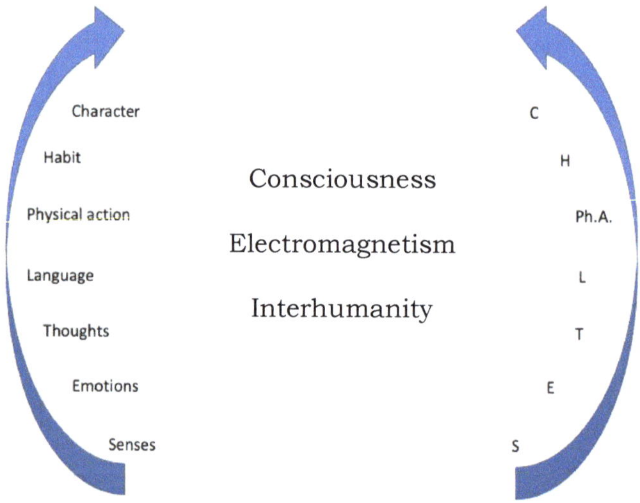

Figure 4. Model of the chain of behaviour

Figure 4 illustrates how **the chain of behaviour** is constituted - **from consciousness, the electromagnetic interhumanity gap - to our senses, emotions, thoughts, language, physical actions, habits and character and further back to the space to be given to others**. Note how it follows the embryological order of priority (chapter 1). Imagine the loop of the chain of behaviour as a horizontal figure eight as a symbol of how the invisible moves between us. Also note the loop which occurs when reading scriptures and learning from digital devices, chapter 22 on Learning.

The chain of behaviour concretizes the abstract consciousness (electromagnetism, interhumanity) **through our behaviours, where we can distinguish between what feels most healthy through our senses and emotions**. The curved arrows on each side of the gap in figure 4, symbolize our bodies as a parabola that both can emit and obtain information. Again note how the behaviours always are linked through stimuli - response in some way.

Communication at the different levels does not have to be, and far from always is, straight lines between the levels (eg feeling to feeling). For example, we can say something that triggers or dampens a feeling, or reinforces or weakens someone's physical actions, habit or character.

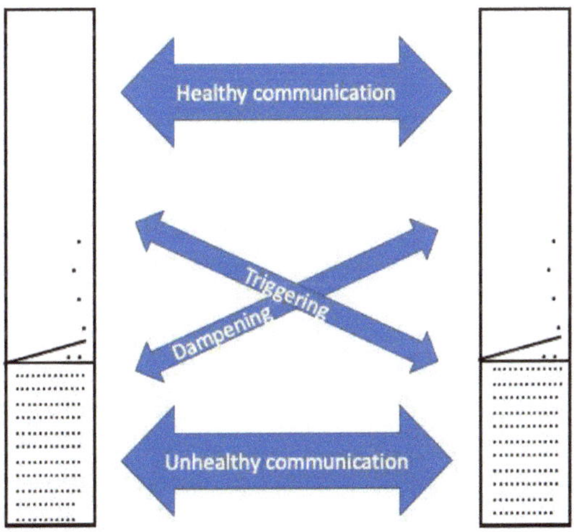

Figure 5. Model of the communication between humans (columns).

The healthy communication (figure 5) intends to symbolize conscious communication and the unhealthy communication (figure 5) unconscious communication (see also chapter 4 Awareness and unawareness). This means that the conscious communication follows the origin of the fractal symmetry that wishes each other well. We want to understand each other, listen carefully and consider what others convey before we respond. The unconscious communication is driven by latent micro-trauma based on emotional blockages and is the inverse communication that has difficulties to receive what others want to convey and just wants to get their own self-centered opinion conveyed. The combination of conscious and unconscious communication (arrows triggering, dampening figure 5) refers to what happens when people with different awareness are communicating with each other. Here we switch between the conscious and the unconscious part, which we (usually) do not have time to reflect on in the moment, unless we're made aware of the fractal symmetry.

Further on we now see how this book time and again wants to show us that something must be given in order to be received. It is not possible to receive anything if nothing has been given. The senses cannot get any information if there is no information available. The chain of behaviour together with other parts of the book, such as the fractal symmetry and the embryological order of priority, thus become self evident.

As an example, think about what it would be like if we put a child in an isolated dark room for the first years of their life. First, it **feels** very unpleasant and cynical (which shows the inherent empathic ability), but the purpose here is to show that there is no information for the child to receive in that situation. Thus, the learning from the surroundings (programming of fractal symmetry) is not possible and the child does not receive any information/lessons for coping with life.

A rule of thumb to remember in all communication towards strengthening the healthy: About 6 seconds before we express something, it is already clear to us what to say. This means that if we respond within 6 seconds after someone finished talking, we have not been able to receive all the information from this person according to the chain of behaviour. Instead, in those cases, to different extent, we have just been waiting for an opportunity to convey our own information.

* Consider the following: Where does the information come from when you are alone? For example, a new idea (genius flash)? Reflect and discuss using figures 4 and 5 above (also take chapter 11 Micro-trauma figure 3 and chapter 9 Sources of information into account).

Note of extra importance: This chapter generalizes the principle of how information wanders through the chain of behaviour. In order to fully understand the exchange of information, this chapter needs to be interwoven extra carefully with the comprehension from chapter 20 Male and Female. Also consider chapter 8 for this note.

15

~ Behaviours are collective ~

No behaviours can be individual and we can draw this conclusion from the reasoning this book drives as a whole. And in the introductory chapter on the embryological order of priority, we actually get this basic understanding from the start and we now fill a bit extra (also remember the reasoning in other chapters).

Behaviours are created from social contexts (see also the principle of The chain of behaviour chapter 14), where we seem to be individuals with different inputs in a specific context. Then we should keep in mind, that a given input either comes from previous input into each individual or through the social interaction in the present moment, or a combination of those (see also chapter 5 Interhumanity).

The individualization we see all around us, ie what disadvantages healthy relationships, stems mainly from what is dealt with in chapter 26 on cooperation and competition (last paragraph). But, at a closer look at the individualization, we can see that each individual is shaped from the outside throughout the fractal chain, which gives us further nourishment to the heading of this chapter.

We can also express it as if the individualization comes from an internalized exterior of an unhealthy nature.

This view of our behaviours, also brings the understanding that everyone has done and does what everyone can and has learned. Based on that everyone is doing or did exactly what they can or could, it thus becomes difficult to judge anyone. However, that does not mean that all behaviours are healthy, but that we should be able to distinguish between what is internalized in a healthy or unhealthy way and which behaviours are and unhealthy based on it - without judging each other. And instead of reprimanding the reprehensible, we should pay tribute to and reward to the desirable.

This can also be compared to what happened when the scientists in The double slit experiment expected a certain outcome (chapter 6 Focus). They were judgmental and influenced what they judged. If they instead had been neutral in their attitude to the experiment, the pattern of the molecular behaviour could have been what it was intended to be in the prevailing situation. It is the same way with us humans. If we judge each other, we lock each other in into specific states. But if we are open and non-judgmental to each other, we are given the opportunity to become what we are supposed to be in different situations.

* If necessary, repeat chapter 12 What is a behaviour?, chapter 13 How does a behaviour arise, and chapter 14 The chain of behaviour.

16

~ Emotions - What's invisible, but is felt ~

We are limited in how to describe where the boundary between, for example senses and emotions is, but when theorising we need to categorize. All parts of our bodies are brilliant and ingeniously interconnected in the fractal symmetry, without waterproof compartments between the different areas.

This theory defines **emotions** as; **what's invisible, but is felt**. For example, it may be joint pain in a knee, muscle pain, stomach pain or feelings of security/insecurity. The former is easier to understand. The latter is more abstract in its character and is what we are primarily dealing with in this book. Emotions such as pain can sometimes be the result of a physical injury from an external force, but in other cases be a symptom of any internal cause (defect) in the fractal symmetry, arisen from unprocessed emotional blockages.

Abstract feelings such as insecurity, well being, irritation, anger, joy, contempt, sadness, surprise and all other words we have for our feelings, are often associated with the surrounding environment and situation, but these can also appear as an effect of emotional blockages being triggered by the environment or during processing.

Emotions that arise from a situation or from the surroundings (presence) can be seen as our reading of the senses of the environment to determine if they are favorable and in parity with our internal original healthy fractal symmetrical core. If so, inner security is brought up and if not, our emotions signals that we are in an unhealthy environment (cf. chapter 1, third paragraph).

It may also be that unprocessed emotional blockages are triggered by a certain environment and give rise to emotions that can be associated with the external environment in the belief that it is the one that is unhealthy, when it in is fact the inner blockage that stands for unhealthiness. This is the case when we are unaware of what blockages we have (see also chapter 4 Awareness and unawareness). In principle, we can express it as; we are in a virtual reality in those situations, where the emotional blockages create an unconscious projection, where we do not really see the person in front of us or feel the prevailing environment. Often incentives for interpersonal conflicts, whether it is within or between the sexes (cf. chapter 14 The chain of behaviour).

Note: Emotions affect the male and the female in different ways. The woman has the fractal need to express her feelings to get support from the man, while the man has the fractal capacity to take care of his own feelings (see also chapters 11, 20, 31).

17

~ Self-reflection ~

When we self-reflect, healthy foundations are built for the strengthening of relationships. In all types of conflicts, we project some of our unprocessed blockages towards each other, which we avoid in self-reflection. Instead, we then reflect on how we reacted/acted in a specific situation, how we felt, if we projected something and if we can do something different for more healthy behaviours at another time. In self-reflection, we also give ourselves the opportunity to process emotional blockages (see also chapter 31 Process and manage).

18

~ Stress ~

At a cellular level, what we call stress, arises when external signals from the outer casing do not harmonize with the nucleus of the cell and do not allow them to pass the casing, in order to protect the core. Based on fractal structure, we can here draw parallels to insecurity. When the surroundings of the cell cannot meet the need for its security, the cell becomes insecure. In the same way, a person becomes insecure, ie stressed, if its external environment feels hostile and does not harmonize in accordance with the inner fractal structure. As a protector, the man is originally better equipped to be able to handle insecure environments, ie stress (see also chapter 20 Male and Female).

Stress also occurs if the environment is healthy, but that where there are unprocessed emotional blockages that are triggered by something in the environment. In these cases, a lack of trust in the environment has been created as emotional blockages, by previous environments being hostile and insecure. Unconscious emotional blockages should always be taken into account when the environment includes other people with unconscious insecurity (see also chapter 11 Micro-trama).

We can find further explanations on stress in the strength of the congruence between the brain and the emotion/senses. The level of stress is then dependent on the difference between these in the fractal structure. That is, the intellect retrieves information from a different time in the fractal chain than what is presented by presence (see also examples 1 and 2 below, chapter 4, fourth paragraph and chapter 29 third paragraph).

The findings described above, activates the bodily functions for fight or flight. We do not deal with the meaning of fight or flight in depth here, but instead refer to what is available in other places about what these bodily functions implies. However, we should note that the mechanisms for fight or flight arise from the micro-trauma and exist as unconscious latent defects in the fractal symmetry (latent overactivity of the sympathy nervous system), which in turn acts as a foundation for unhealthy behaviours. This is of great importance for the purpose of the book and for the understanding of the importance of processing emotional blockages (if necessary) in favor of healthy behaviours.

Of additional importance, is to understand that stress, ie fight or flight state, to varying degrees (depending on strength) turns off the contact with our senses and emotions in real time. This means, that during periods of stress, we can only care about ourselves, since the principle is based on protecting ourselves from external danger (cf. being chased by predators). Which condition generates unhealthy behaviours because the stress we primarily are talking about here, is based on emotional blockages from micro-trauman and in those moments it creates a virtual reality.

Breathing is what is most relevant to the context of fight or flight and it is also an direct indicator of harmony/disharmony between the internal and the external at all levels in the fractal structure, ie when the fight or flight functions are activated or deactivated with breathing as a common denominator.

Fig. 6. Stress/Disharmony
Unstable levels of energy

Fig. 7. Safety/Harmony
Stable levels of energy

When talking about stress/disharmony in terms of frequencies, we can see it as a high frequency phenomenon, figure 6, and security/harmony as a low frequency phenomenon, figure 7. These phenomena can be verified in practice, by how fast and shallow breathing arises when there is some kind of disharmony and how calm and deep breathing is found in harmonious contexts.

Some examples of how stress occurs:

1. We are packing a bag before going to meet some friends for a weekend and we experience an internal stress, but do not know why. During a previous therapy session, we found that there was a childhood strategy to flee from social contexts because the environment was insecure. In this example, the resulting strategy due to the unprocessed emotional block-ages (micro-trauma) is to flee from social contexts. This is what gives rise to the stress because we are on our way to a context that is diametrically opposite from the learned strategy. Stress then arises because thought and feeling (brain and body) are incongruent.

2. An oral presentation by all students is announced by the teacher and suddenly the rhythm of the heart increases. A conscious or unconscious trauma is being triggered and we lose presence and the congruence between emotion and brain deteriorates significantly. In our early years we may have been exposed to laughter or mockery in any or many situations when we spoke in front of, for example, the family or in family situations. Often during silence that was perceived as troublesome, because normally people were chattering constantly. When we later in life find ourself being in focus in a situation where there is silence, the unprocessed emotional blockages will appear and give rise to internal reactions of stress (fight or flight).

3. You are hungry and feel stressed and get a feeling that you need food immediately. If you for example at some point have been misunderstood by your mother during infancy and she did not understand that you were hungry (cf. micro-trauma), then stress was created around you not getting food when needed. Later in adulthood, this blockage may create stress even though there is food available and the environment is supportive.

Food and drink (nutrition) overall are big parts of life and can create stress or create the desire for similar frequencies. For example, if we are stressed, it is easier to turn to rapid carbohydrates that to harmonize with the stress and vice versa, it can trigger stress. It is for this reason that a low carbohydrate diet is good for internal health and healthy behaviours.

19

~ Relationship as a need ~

This does not mean that we are primarily dependent on a relationship to get something from it, but because the love within becomes static and creates illness if we are not allowed to cooperate and to share what is meant to pass trough our bodies (according to the chain of behaviour chapter 14 and according to the principles of male and female chapter 20). The desire to share of oneself follows the fractal principle and if followed by both parties in a relationship, then both will get what they need. As much as possible, depending on the state of the original fractal symmetry.

This reasoning is based upon the fact that we are constituted to be both giving and receiving. **Give and receive** (instead of the saying "to give and to take") in accordance with chapter 28 Unconditional love (see also chapter 20 Male and Female).

The existence of codependency stems from what is described above, but becomes a consequence of residual micro-trauma and a feeling that the other does not give or receive what we need to give or receive. Our strategy then becomes to be demanding, rather than giving and receiving (fractally inverted) and the relationship turns unhealthy. To various degrees, we will also become more self centered, than to care about the other (unconscious levels), despite the fact that we are created to give and receive. See also the next chapter, Male and Female.

* Reflect and discuss the truth of the following: **Give** what you want and you will **get** what you **need**.

20

~ Male and Female ~

We are all humans, but we are also men and women. Why do men and women exist if we should just see ourselves as equal people? Of course we shouldn't, but instead it implies our different purposes as humans. A male purpose and a female purpose, which in a unit provides a common purpose. We need to recognise and affirm important fractal symmetrical biological differences for the compatibility that men and women represents for the unity.

We are fractally structured from the invisible (electromagnetic fields, the psyche) to the visible (visible fractal structure, the body). This means that the same fractal symmetry that governs our psychological characteristics also governs our bodies. This in turn means, understanding what the male and female bodies are intended for, we also understand what their respective behaviours, communication, needs and psyche are intended for. And it is enough to understand one principle to understand the whole. It lies in the understanding of the nature of fractal symmetry. For that reason, it is easy to proceed from the state of reproduction.

It further implies that **the man is a clear giver and that the woman is a clear receiver**. Remember here, too, the introductory reasoning in the first chapter (Embryological order of priority). And for a man to be able to give something, it needs to be gathered. And what a man then gives to a woman, will be received when she gives herself (invites him) to the man in order to receive. And again, **the giving is completed only when the woman is ready to receive** (Completion also means mutual acceptance of each other. Cf. chapter 1 last paragraph).

The ability of a woman involves multiplying and refining what is received, to **give it back** with more life and in a refined form.

Then the man protects what has been refined and given back and provides what is needed for the continued cooperation between the two. As interwoven giver and receiver for the continued generation of the fractal chain. This principle can then be applied at all levels in terms of behaviour, communication, needs and psyche for both the male and the female in agreement with the constitution of the fractal structure. Note the bold text; that **the giving from both is central to the receiving, even if the receiving is just as important as the giving** (see also chapter 28 Unconditional love). Also note that the giving by a woman is based upon her inviting, receiving and refining.

The masculine and the feminine are constituted to help each other manifest all that they share together with this intertwined and collaborative principle. None of them could do this work of unification on their own.

In other words, the man and the woman respectively are intended to do what the other is not intended to do and it is important to emphasize the equal value of men and women and that men and women are equally important in their qualities of compatibility for their common purpose as a unit. Man and woman must be seen as a unit, which will have the needs for both of them fulfilled for their purpose in a close love relationship.

Following the principles of masculinity and the femininity gives us basic fractal benefits for healthy relationships.

Below we put some words on the differences between men and women for further consideration, reflection and discussion.

Man - Provider, leader, leading responsibility, work, protection, security, entrepreneur, provider of materials, giver, developer/ nurturer of the masculine. The man is the main giver and thus has the main responsibility for the family. The man is equipped to take care of his own feelings in order to provide the family with security.

Woman - Receiver, helper and developer of the received, refinement, caretaker, encourager, incubator, breeder/nurturer of the feminine, resource. The woman is the main recipient and is responsible for what is received. The woman receives help with her feelings from her husband as the recipient of his security.

Both the man and the woman need to **feel** that their giving is received through actions of the entire chain of behaviour. They are also equipped with qualities to strengthen each other's value in the fractal symmetry (restore if necessary). See also chapter 28 Unconditional love.

Many misunderstood interactions between men and women in our world are due to a forgotten understanding of what male and female means in practice. If a relationship is unhealthy, it is on a deeper level a symptom that something is defective in the fractal symmetry, ie something that can be derived in the fractal chain from the point where unhealthy information changed the fractal symmetry. As long as the defect persists, symptoms of unhealthiness will persist and/or appear in a different form of unhealthiness in the relationship.

Metaphor: If we bake a cake and one of the eggs was rotten, the whole cake would be affected by the rotten egg, Similarly, if there is a defect in the male or female fractal symmetry, then the relationship will deviate in some way or another from how it was created to be. Symptoms in the relationship due to some cause in the fractal symmetry.

Note: If the principles this chapter is providing, are inverted, then we end up in the field of structures characterized by what is explained about narcissism in chapter 34 Empathy and narcissism. Otherwise see also chapters 11, 16, 31.

* Reflect and discuss.

21

~ Language is not a sense ~

But, an expression of the sensual. Depending on the state of the fractal symmetry, the language can in various degrees convey what the presence conveys to the senses. Alternatively, what emotional blockages convey from previous micro-trauma. Or a combination of both.

Since language fractally comes after senses and emotions in the natural order, it is automatically colored by what these convey. Both in terms of what is in the presence and what is stored in the fractal symmetry (as defects).

That the language is not a sense becomes self evident by the fact that what was received by the body (the senses) now comes out in a different form to be given to others (see also chapter 14 The chain of behaviour). This is because senses cannot give to themselves as they are first in the fractal structure. As a comparison, remember that the brain is an organ, that with help from sensory impulses, is involved in creating the language.

Concepts used by the language can be pronounced with different frequencies and through this, the truthfulness of what is conveyed can be understod. It depends on what the division of the clean part and the micro-trauma in the body look like (cf. figure 3 chapter 11 Micro-trauma). Regardless of if we are aware of our micro-traumas or not, they ultimately are a part of creating our language. In order to be able to read the truth, it is required that one's own emotional blockages at the current level are released.

Through the metaphor of how we tune a musical instrument to make it sound clean, we can draw parallels to what is needed to get a clean language with healthy frequencies. That is, maintain or restore the fractal symmetry.

22

~ Learning ~

The curiosity of the fractal structured giving, from the moment we come in to this world, makes us constantly strive to learn. The question is how do we best learn something? Learning, to begin with, is fundamentally synonymous with programming. We bring something into our bodies to be stored for use in various contexts later in life.

Following the chain of behaviour during learning, based on the fractal symmetry and the embryological order of priority, it will be **presence based on the experiences of the senses**. Every level of the fractal programming becomes then healthily reinforced by the actual environment during learning.

Commonly, learning involves reading scriptures and learning from digital devices, which means we will skip the first two steps in the chain of behaviour, because the text in scriptures and digital devices are not a part of the environment of our senses. We use vision, which certainly is a sense, but the text comes from an artificial environment that is not connected to presence, instead comes from another place when the text was written down by whoever wrote it at that time. In the best case, the author was connected to the presence and wrote based on it, but the most common is writing directly from the intellect (brain). We will then loop without senses and emotions in the chain of behaviour and get unhealthy programming in the fractal symmetry.

In the absolute presence, where the senses and the body are, there are no words and no concepts (remember the child's communication before learning the language). Words and concepts are something we shape with the help of the brain to describe the absolute presence. This means that words and concepts, more or less, always will be distanced from the presence, which is beyond the conceptual sphere. This, in turn, means that words and concepts have subtle differences in terms of interpretation and are dependent on how and what we learned about them. Our perception is put out of play to various degrees when only learning through reading scriptures and learning from digital devices.

Basically, we can learn anything through reading scriptures and learning from digital devices, without knowing whether it is true or not. And without reflection, it may very well be that we walk around with various amounts of untruths stored in our bodies (see also note chapter 4 last paragraph). In those cases, there is some form of sensory suffocation that misleads our ability to perceive. And that we then convey untruthful information to those around. In other words, learning through reading scriptures and digital devices needs to be verified by the senses to know if what we read is in alignment with the senses and the emotions. For example, if I read that all trees are red, I need to go out among trees and have this verified or denied through the impressions from my senses. Or for that matter if I hear a rumor about someone; then how can I know if it is true or not?

* We conclude this heading with some disruptive questions for reflection and discussion:

- How did I learn what I believe I know?

- How do we learn best? Through the experiences from our own senses or through studying (reading scriptures and learning from digital devices)?

- How do I know if the person I am receiving information from is conveying sense-learned or reading-learned information? What determines this and how do I determine it? And who can learn from whom?

- If I carry a strong belief, that in fact in real life is a lie, of which I am unaware, am I then willing to cooperate with the truth?

23

~ Memory ~

A memory is part of the fractal structure programmed into the fractal symmetry at the time the memory was created. When we think of an old memory, we simply go back to the point/time when it was stored by sense impressions on that specific occasion. It can also be intellectual memories, but these are not anchored in the fractal symmetry in the same way as sense-learned memories (see also above and chapter 22 Learning).

Repressed memories (trauma) remain as encapsulated emotional blockages in the fractal structure as above, but are not consciously accessible until they have been processed (see also chapter 16 Emotions).

The function of the memory has great value in being able to relive how life should be felt, and how it shouldn't. It helps us finding unprocessed blockages in order further strengthen the healthy. Memory is also a large part of the lessons of life.

24

~ What does it mean to think for oneself? ~

This question can be seen as a complement and a further development from chapter 9 on Sources of information. The philosophical question about the free will may be actualized for some readers and it is in a way relevant. But at the same time it has become dichotomous, which in many cases lead us astray and is therefore irrelevant to the truth of the question. We hereby leave that question in this book.

Many, perhaps even all, would probably claim that we think for ourselves. It seems that way when we are lost in thought by ourselves. At the same time, we have now received information from the other chapters that we all are connected in a very intricate and ingenious way. This through the ingenious fractal symmetry, from the inside out (which is actually the same thing, see also chapter 5 Interhumanity).

At a closer look based on what we already learned, when it seems we are thinking for ourselves, the thoughts come from a bottom of micro-trauma (emotional blockages, the dotted field, picture 3, chapter 11 Micro-trauma) and in other cases, they come from the surrounding environment according to the chain of behaviour, alternatively what we learned through various forms of education. Of course, it can also be a combination (and often is), but the bottom line is, that in principle, we cannot claim that we think for ourselves.

See also primarily parallels with chapter 14 The chain of behaviour and chapter 15 Behaviours are collective.

* A chapter for self-reflection and contemplation.

25

~ Reciprocity ~

Reciprocity at its foundation is healthy if healthy behaviours elicit healthy behaviours in others. Reciprocal actions are based upon group belonging being more important than the self. Which also means that the individual has a greater chance belonging to the group (greater chance to survive) if the behaviours benefit the group as a whole and generate health on all levels. The principle of reciprocity is to behave towards others in the same way that I want others to behave towards myself and others in the group - The Golden Rule.

One can also talk about reversed reciprocity. Which means an unhealthy affiliation to the group. For instance, a group based on thinking the same about someone else (something else) even if it is not true. And to choose an excluding position in that belonging. Bullying and psychological abuse are examples of reverse reciprocity. The principle of reverse reciprocity is based on doing to others what I myself fear being exposed to, in order to avoid it for my own benefit - The Golden Gule inverted.

Compare with chapter 34 on Empathy and narcissism.

26

~ Cooperation and competition ~

The idea that cooperation is healthier than competition, stems primarily from deepened understanding of the fractal symmetry, but also from what inter human experiences tell us (higher up in the fractal structure).

The concept of symmetry tells us a lot about alignment, giving, a common endeavor and wishing each other well. Emotionally, it also harmonizes with what cooperation implies in its essence. Namely, helping each other to develop or improve something, which also provides inner satisfaction. At the cellular level (further down in the fractal structure) cooperation and fractal symmetry are verified by cooperating electromagnetic fields reinforcing each other and thus tend to wish each other well. Cooperation thus benefits the collective, from which all behaviour origin-ates.

Competition is often said to favor development in a societal context. But, how does it affect the individual (fractal cellular)? First of all, the concept of competition. It means doing something better than someone else at the individual level or at the cellular level. It then becomes a quest to want to be better than others, a kind of competition where the winner is rewarded. Unlike collaboration where the collective is unconditionally rewarded. It may feel good to the winner with competition, but is it fractally symmetrical and behaviourally healthy? At the fractal cellular level, we can understand that the cells will not cooperate and strengthen each other, leading to weakened electromagnetic fields for the body as a whole and with other people. It thus gives an unhealthy fractal symmetry as well as unhealthy behaviours.

The context of sporting competition has another perspective on com-petition that gives us an opportunity to understand the hierarchy that competition creates. A hierarchy that is based on being more respected the higher up you go, which implicitly also means that those below are not worth as much as the winner. Cooperation, on the other hand, has no hierarchy in that sense, but is instead based on everyone's equal value and gratitude for everyones contribution, regardless of the task being performed.

Cooperation is thus what this theory highlights as healthy behaviour.

27

~ Codependency ~

Here, codependency is defined as **a substitute for unconditional love** (chapter 28). Codependency consists of the demand by defects in the fractal symmetry that strive to want something instead of giving. In relationships, this is shown by the fact that interacting micro-traumas give rise to structures in behaviour that often blame the other for their own unprocessed micro-traumas. Alternatively, trying to fill the other with what he or she requires without being able to receive because of their unprocessed micro-trauma (more detailed in Chapter 34 Empathy and narcissism). Expressing it in another way; wanting something healthy in the other that we don't have in ourselves due to unhealthy changes in fractal symmetry.

Codependency in this definition is unhealthy. The foundation for its existence stems from what is described as a need in chapter 19 Relationship as a need. It is therefore important to be able to distinguish between what is unhealthy codependency in relationships and what is healthy dependency in relationships, for example with regard to male and female complementary characteristics (See also chapter 20 Male and Female). The latter particularly evident in a close love relationship.

28

~ Unconditional love ~

The heading is a concept becoming more and more common and it needs to be understood from the embryological order of priority, the fractal symmetry and the chain of behaviour. Altruism is an entrance to understanding the attitude and approach toward oneself in such a state as this book describes unconditional love.

The definition of the concept of unconditional love then means that **giving and receiving** must take place without conditions and defects at all levels, as well as follows the principles from which we were created. If that happens in a relationship, that relationship will be perfectly healthy and its behaviours will mutually follow anything that continually improves the relationship.

Giving, so to speak, is the fractal keynote for unconditional love, which then gives different expressions for the purpose of men and women as a unit (see also chapter 20 Male and Female third paragraph).

In Chapter 1 on embryological order of priority, we learned that the sperm was let into the egg. This means that the sperm **was** first **given** to the egg that received it. The sperm **was** then **given** access to enter the egg. And to nuance it further, it means that nothing can be received if nothing is given and nothing can be given if there is no receiver. We can also draw parallels to healthy cells, where these are open to both giving and receiving when their respective inputs (giving) to each other are in tune with their inner core (receiving). It's the same way with unconditional love. There are constantly ongoing intertwined acts of giving and receiving. This is particularly clear in a love relationship between a man and a woman (see also chapter 20 Male and Female and chapter 14 The chain of behaviour).

We always strive for this unconditional state on the conscious level and everyone does their utmost to be in this state. That's how fractal symmetry works. In different parts of this book, we describe what can prevent this optimally healthy state.

29

~ Heart, emotion, brain ~

Back to the embryological order of priority (chapter 1) where the mother from the beginning provides all information from her blood to the heart of the embryo. It gives us the understanding that the heart can receive all information from outside (embryo) and also send all the information that is in our bodies (mother in this example) (cf. chapter 9 Sources of information and chapter 14 The chain of behaviour).

Furthermore, based on the embryological order of priority, the fractal symmetry and the chain of behaviour, we are here again reminded that all information pass our bodies through all described parts and finally reach the brain. Which is secondary to the body, but extremely important for processing the information from the body in the appropriate way (without external influences such as reading scriptures and learning from digital devices).

The simplification in the heading of this chapter implies that the heart is central because it came first (without taking breathing into account at the conception stage). Here, we are talking about the wide array of the frequencies that can be received and transmitted.

The information through the emotional life (number two in the embryological order of priority) is finally translated/interpreted by the brain. This means; that if the healthy information unadulterated flows throughout the chain, these three parts, heart, emotion, brain, will be in congruence and in optimal health. The interacting electromagnetic fields of the body is strengthened and the field around such a person becomes healthy for others as well. The strength of the electromagnetic field we all carry depends on the state of the fractal symmetry mainly through these parts.

30

~ Needs and cravings ~

Defects in the fractal symmetry give rise to various desires such as craving for sweets. Sweets are not a need from the body, but for various reasons the brain can become addicted to fast carbohydrates. It then wants fast energy directly from the food intake instead of the energy getting through our internal organs and then get to the brain. It can also be seen as a form of stress, ie that we have gotten used to not waiting for the natural process of the body when we get the nutrition it is built for. Cravings can also be, for example: Too much exercise, excessive attraction to sex, excessive surfing, identification with things, titles, etc.

The needs always follow the fractal chain of the embryological process and if we keep it intact, or recreate it, we will preserve our inner security and our healthy calmness. Needs are, for example: Sleep, species-specific nutrition, clean water, social security, being included and feeling loved, sun, clean water, fresh air, etc.

Needs and cravings need to be able to be distinguished for processing and managing (chapter 31).

* Reflect and discuss.

31

~ Process and manage ~

A few words about the difference between processing emotional blockages and dealing with the same. Managing emotions in this context means putting them on hold for a while (putting a lid on the emotions) and to do something else. The processing is what it sounds like and means to really meet the emotions in the body and have them released in order to get access to the information contained in the blockages from different traumas.

The emotional blockages in the body are powerful and static. Having them released will be feelt in one way or another. Some are short-term and volatile while others are persistent and lasting in their state. It's important to have the knowledge that everything that is experienced and changes is for the good when processing. In principle, it is like stretching. If we stand in a certain position to extend a muscle, the tension will be released after a while. The principle is the same when we process our emotions. When we are in the emotions, we will feel how they change. Then we have done or started the processing of a certain blockage.

Spider phobia is another principle metaphor. In spider phobia, the individual needs to slowly approach the spider in order to eventually be free from his fear of spiders. On your own or with the help from someone. It is the same with emotions. We need to approach them gradually and sometimes we process on our own and sometimes we need support with the process. In this example, we can also see that emotional blockages are what give rise to fears. And if we are scared, we cannot care for others, which is unhealthy (ch. stress).

Processing emotional blockages takes energy and we can't do it all the time. Therefore, it is important to also have strategies for dealing with these. Preferably in nourishing social contexts with joy and fun activities. But, it is not always we want that and then of course it is fine to do something on our own.

A key to accessing our unconscious emotional blockages comes from understanding our strategies that will avoid them. These were generated during our early years of life and are still active if they are not processed. These are often very easy to see (feel) from the outside for those who have learned to see (feel) them, but much more difficult to discover on our own, just because they are unconsciously blocked.

Feel free to think in terms that the different parts in the body are like tuning forks for instruments that need to tune down (release emotional blockages) for more and more fruitful behaviours and for improved fractal health.

Fasting is a great tool for accessing emotional blockages for processing. Fasting will help us to recognize self centered unhealthy strategies and find where they are funded. Fasting (depending on length) also get autophagia started, which is a way to describe repair of the fractal symmetry.

Note: Processing and managing is different for the male and the female. The man has the capacity to take care of it himself, while the woman is not equipped for it in the same way (see also chapters 11, 16, 20).

32

~ Dynamic and static ~

The healthy structure of the fractal chain, ie life, is in its essence dynamic and when damaged, static emotional blockages and static physical blockages are created, which generate ill health and unhealthy behaviours. Dynamic structures are thus what create healthy behaviours.

These two concepts can also be seen as examples of how what we repeatedly are learning, is hiding behind different concepts. Concepts that we use to describe what's beyond them - from where everything arises.

33

~ Mental (illness) health ~

Mental (illness) health with illness in parentheses is a symbol of the importance of **focusing on health** (see also chapter 6 Focus). At the same time, what we call mental illness, is actually a description of when we have gotten in contact with emotional blockages at different levels in the fractal symmetry and are in various conditions on our way to internal recovery (healing) of the fractal symmetry. If the person is aware of the fractal symmetry and knows how to proceed to process these blockages (or getting help with them), it is the way towards more healthy behaviours.

Anyone who has gotten stuck in an identification to a certain state (heavy emotional blockages) will find it harder to restore their fractal symmetry even with the help from someone. But, it is possible.

Regardless of what the defective symmetry looks like, all different conditions must be seen as temporary and possible to change in order for them to be possible to change (cf. again chapter 6 Focus).

Many are likely to recognize the conditions we refer to as depression and burnout. In this context, these conditions can serve as examples of how we feel when we do not follow the natural fractal principles, primarily in accordance with chapters 1, 2, and 14.

34

~ Empathy and narcissism ~

What we call **empathy** is here defined as; **the ability to feel with other people and to have compassion for others**. Not to confuse with unhealthy sympathy that leads to sustained suffering. Genuine empathy helps the other to release their emotional blockages through their ability to feel the feelings of others. This is only possible for persons who have processed their own emotional blockages at the current fractal level. The more processed the emotional blockages, the deeper the empathy becomes and follows the healthy fractal symmetry.

The quest to always see and exist for others is the foundations for the empathic condition.

The conditions we have come to call narcissism and psychopathy are the diametric opposites to empathy and are here defined as; to be unable to feel their own and other's feelings to varying degrees. These strong fractal defects have their foundation in very unfortunate circumstances in the formation of micro-trauma, which subsequently results in such severe emotional blockages that emotional life is almost completely inaccessible in the most difficult cases. This can also be latent for many years until the person meets someone who triggers these emotional blockages, whereby the conditions for unhealthy behaviours arise.

In the condition of narcissism and psychopathy, the fractal symmetry is inverted. That is, the ability to follow our innermost nature of interacting in a healthy way with other people, is to varying degrees is set out of play. And it leads to extremely unhealthy behaviours for themselves and their surroundings. It is synony-mous with constantly living in the state of fight or flight without being aware of it and at the same time living with the belief of being empathetic.

Of importance to this heading, is the difference between intellectual believing in empathetic ability without any significant emotional processing and genuine empathetic ability as above. And being able to distinguish between these for increasingly healthy optimized behaviours (see also chapter 8 Healthy/unhealthy behaviours and chapter 17 self-reflection).

The ability of the environment to determine what is intellectual versus genuine empathetic ability in a specific individual, is also determined by how well a specific person managed to process the emotional blockages at the current fractal level. That is, how conscious the actual person is (see chapter 4 Awareness and unawareness).

~ Summary ~

We have now gone through the visible and invisible fractal reality from the embryo to who we are today, by putting words into what we cannot see, but which exists and can be felt. The theory claims that our behaviours exist in the way we have now learned and through concretization we can simplify the view of how interhumanity works and interacts in all different meetings between people. Furthermore, we can more easily understand how and why we behave as we do and what we all can do to strengthen or improve healthy behaviours. For the well-being of man and society.

Understanding how the fractal principle includes rewarding, collaborative, inclusive, curious, health promotion and health prevention functions, has been deepened by the chapters and it also implies, the understanding that it includes a self destroying function for the unhealthy when adding healthy frequencies, since then the unhealthy doesn't get nutrition/focus. A kind of self-organizing win win so to speak.

The outcome of our behaviours is always observable and this means that we can relate to each other with a new kind of lifestyle. A lifestyle where we are constantly vigilant to hone our ability to be fellow human beings from the fractal health perspective. An approach we can practice with self-reflection as a foundation to benefit the interpersonal dynamics.

This can be practiced in every everyday situation. And for ourselves, with the senses and emotions, determine what feels right to do to improve relationships and thus heal our own and the common fractal symmetry.

We are each other's fellow human being and any action that strengthens the inter-human unconditional giving, will also contribute to reinforcing the healthy frequencies in our world - For truth and for love.

Stefan Alldén